Nelson Science

This book belongs to:

_____

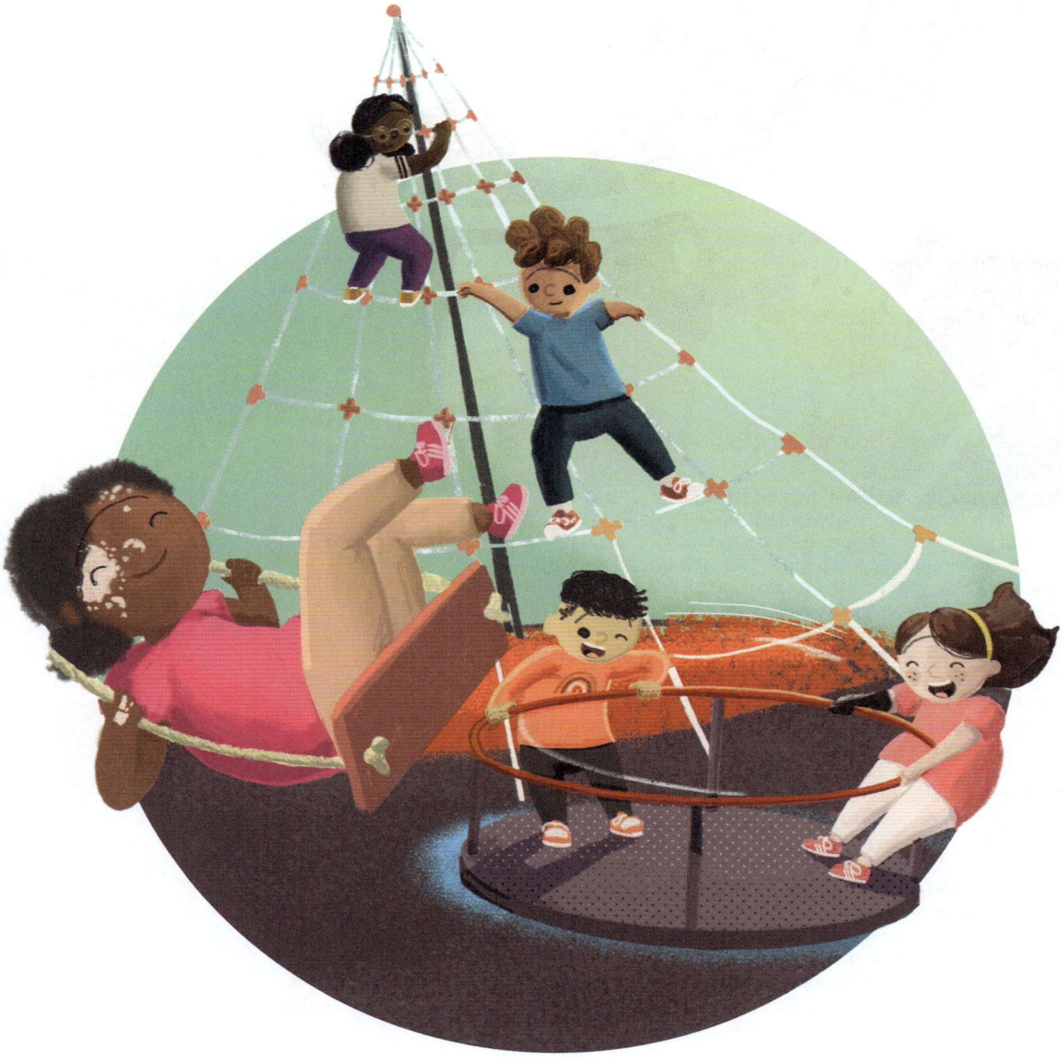

Workbook Starter C

Anthony Russell

OXFORD
UNIVERSITY PRESS

OXFORD
UNIVERSITY PRESS

Great Clarendon Street, Oxford, OX2 6DP, United Kingdom

Oxford University Press is a department of the University of Oxford.

It furthers the University's objective of excellence in research, scholarship, and education by publishing worldwide. Oxford is a registered trade mark of Oxford University Press in the UK and in certain other countries.

Text © Anthony Russell 2022
The moral rights of the author have been asserted.

First published 2022

British Library Cataloguing in Publication Data

Data available

ISBN: 978-1-3820-1780-0
ISBN: 978-1-3820-1781-7 (Workbook only)

1 3 5 7 9 10 8 6 4 2

Paper used in the production of this book is a natural, recyclable product made from wood grown in sustainable forests. The manufacturing process conforms to the environmental regulations of the country of origin.

Printed in Great Britain by Bell and Bain Ltd, Glasgow

Acknowledgements

The publisher and authors would like to thank the following for permission to use photographs and other copyright material:

Cover: Aaron Cushley. **Photos: p7(a):** Jagodka/Shutterstock; **p7(b):** Absolutimages/Shutterstock; **p7(c):** Jan Martin Will/Shutterstock; **p7(d):** Zaichenko Olga/Shutterstock; **p7(e):** Eric Isselee/Shutterstock; **p7(f):** Mariait/Shutterstock; **p7(g):** Roger Clark ARPS BPE1/Shutterstock; **p7(h):** Nikolai Tsvetkov/Shutterstock; **p7(i):** Erik Lam/Shutterstock; **p7(j):** Tsekhmister/Shutterstock; **p7(k):** Pakhnyushcha/Shutterstock; **p7(l):** Eric Isselee/Shutterstock; **p10(a):** Iakov Filimonov/Shutterstock; **p10(b):** Pan demin/Shutterstock; **p10(c):** Alexey Seafarer/Shutterstock; **p10(d):** Roman Samokhin/Shutterstock; **p10(e):** Dmitry Polonskiy/Shutterstock; **p10(f):** Anton_Ivanov/Shutterstock; **p10(g):** VladyslaV Travel photo/Shutterstock; **p10(h):** JaySi/Shutterstock; **p16(a):** Eric Isselee/Shutterstock; **p16(b):** Stepan Bormotov/Shutterstock; **p16(c):** Oleksandr Lytvynenko/Shutterstock; **p16(d):** Steshkin Yevgeniy/Shutterstock; **p16(e):** Tracy Starr/Shutterstock; **p16(f):** Eric Isselee/Shutterstock; **p16(g):** Nikolai Tsvetkov/Shutterstock; **p16(h):** Aaron Amat/Shutterstock; **p16(i):** Marco Uliana/Shutterstock; **p18:** Singkham/Shutterstock; **p21(t):** Spiroview Inc/Shutterstock; **p21(m):** Jessica.kirsh/Shutterstock; **p21(b):** Lopolo/Shutterstock; **p28(a):** ImageBROKER/Alamy Stock Photo; **p28(b):** Granger Historical Picture Archive / Alamy Stock Photo; **p28(c):** Gary Hebding Jr. / Alamy Stock Photo; **p28(d):** Omer Messinger/EPA-EFE/Shutterstock; **p28(e):** Granger Historical Picture Archive / Alamy Stock Photo; **p41(a):** Subbotina Anna/Shutterstock; **p41(b):** Valentyna_Hir/Shutterstock; **p41(b):** Charlie Hutton/Shutterstock; **p41(c):** Sumroeng chinnapan/Shutterstock; **p41(d):** Robyn Mackenzie/Shutterstock; **p41(e):** Dorling Kindersley ltd/Alamy Stock Photo; **p41(f):** photolibrary.com/Getty Images; **p41(g):** R.Filip/Shutterstock; **p41(h):** Mike Stone/Oxford University Press; **p41(i):** D. Callcut/Alamy Stock Photo; **p41(j):** Philip Dubois/Alamy Stock Photo; **p41(k):** Brester Irina/Shutterstock; **p43(tl):** Monkey Business Images/Shutterstock; **p43(tm):** Odua Images/Shutterstock; **p43(bl):** Anton Brehov/Shutterstock; **p43(br):** Mauritius images GmbH/Alamy Stock Photo; **p43(tr):** Maria Sbytova/Shutterstock.

Artwork by Q2A Media Services Pvt. Ltd.

Every effort has been made to contact copyright holders of material reproduced in this book. Any omissions will be rectified in subsequent printings if notice is given to the publisher.

MIX
Paper from responsible sources
FSC® C007785
www.fsc.org

Contents

Plants

1 Count each type of leaf. Write the numbers in the blanks.

_____   _____   _____   _____

2 Collect 3 different leaves.

3 Put your leaves in order of size, smallest to largest.

4 Draw them in order of size.

1 Name the seeds. Draw lines from the names to the seeds.

a)

b)

c)

d)

pea

cashew

sunflower

bean

2 Colour the flowers below.

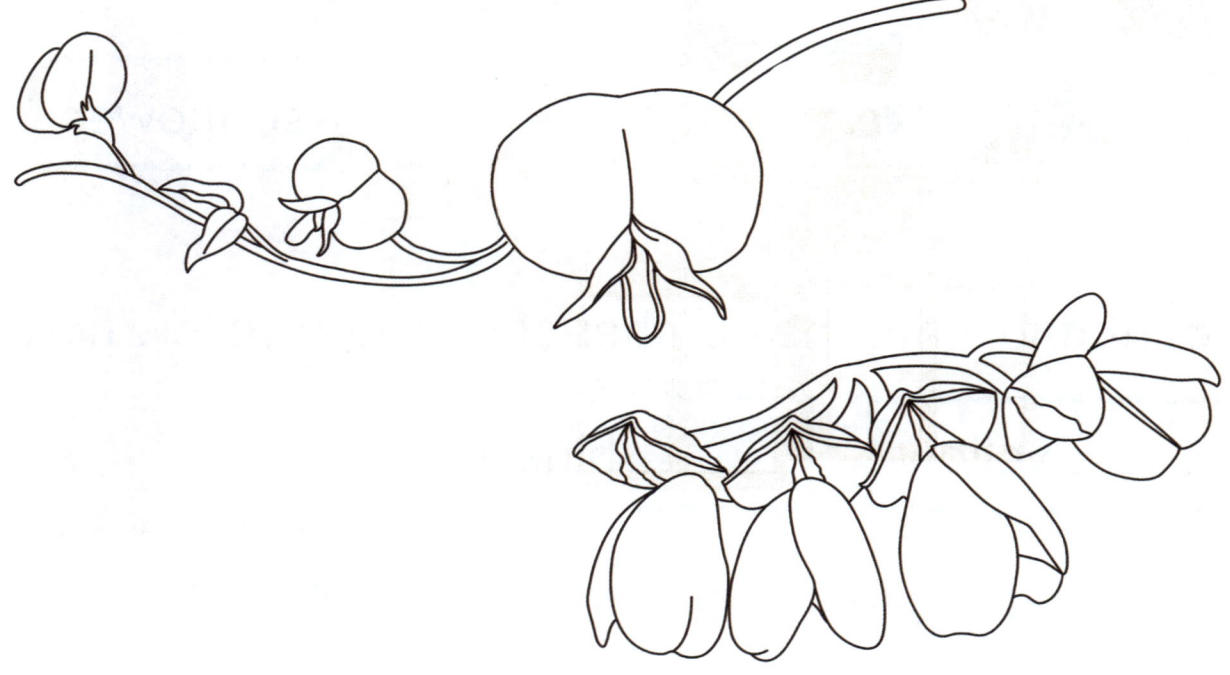

3 Count the red flowers. Write the number in the box below.

_____ red

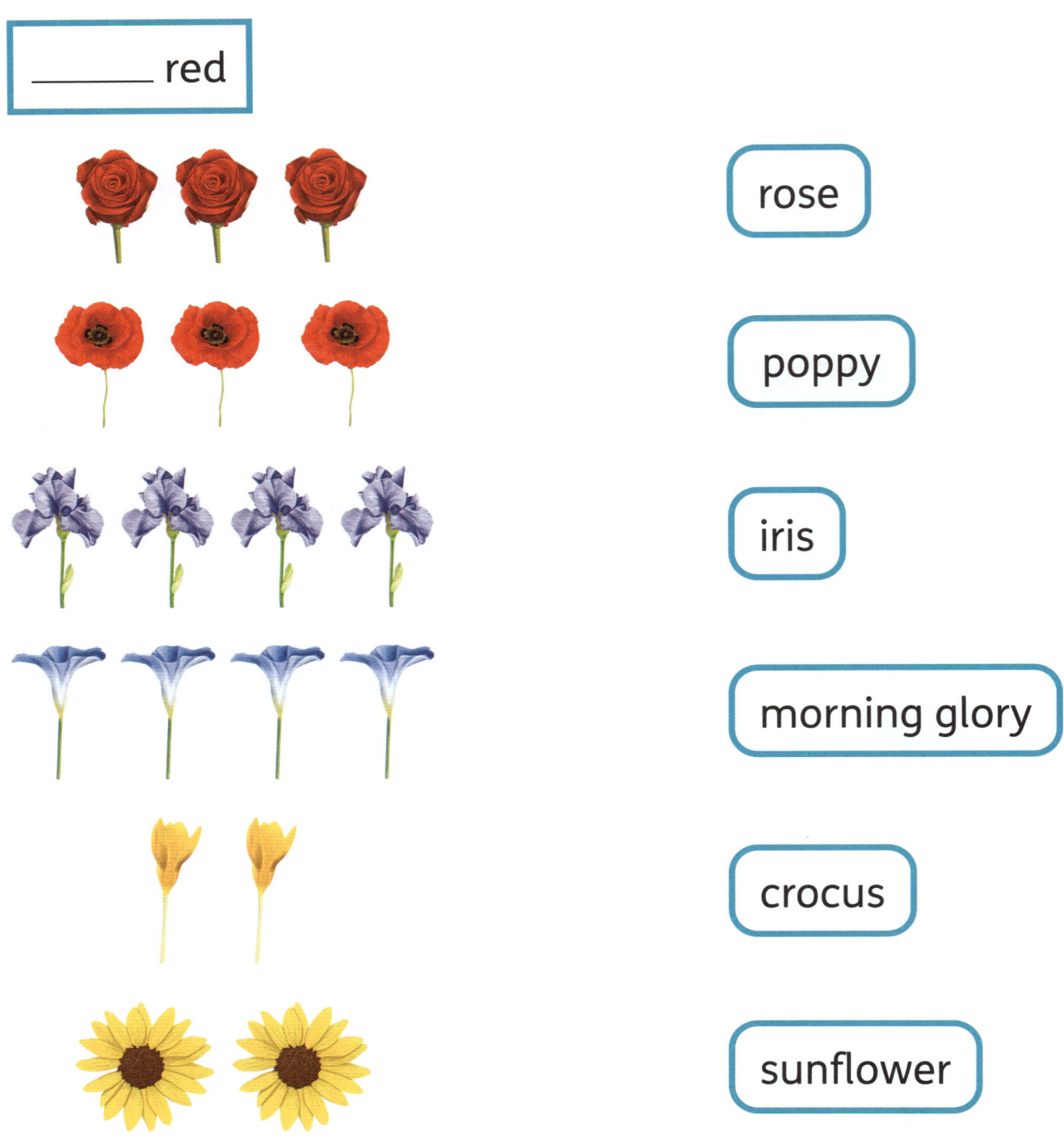

rose

poppy

iris

morning glory

crocus

sunflower

4 Count and write the number of blue and yellow flowers.

_____ yellow   _____ blue

5 Find two flowers with different shapes.

6 Draw them and colour them.
   Show and tell.

Fruits

1 Colour the fruits.

2 Name the fruits. Draw lines from the names to the fruits.

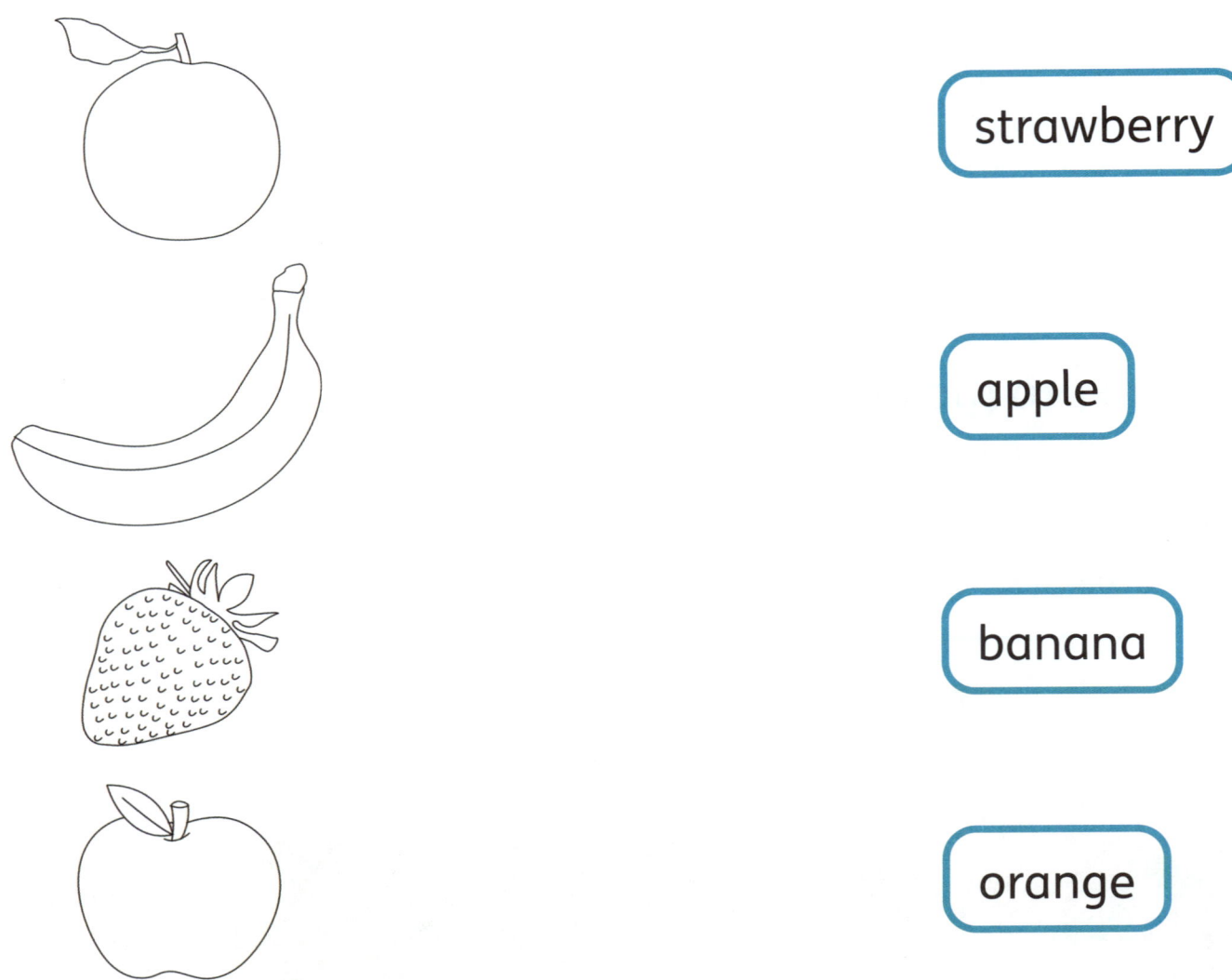

strawberry

apple

banana

orange

3 Draw one flower and one fruit. 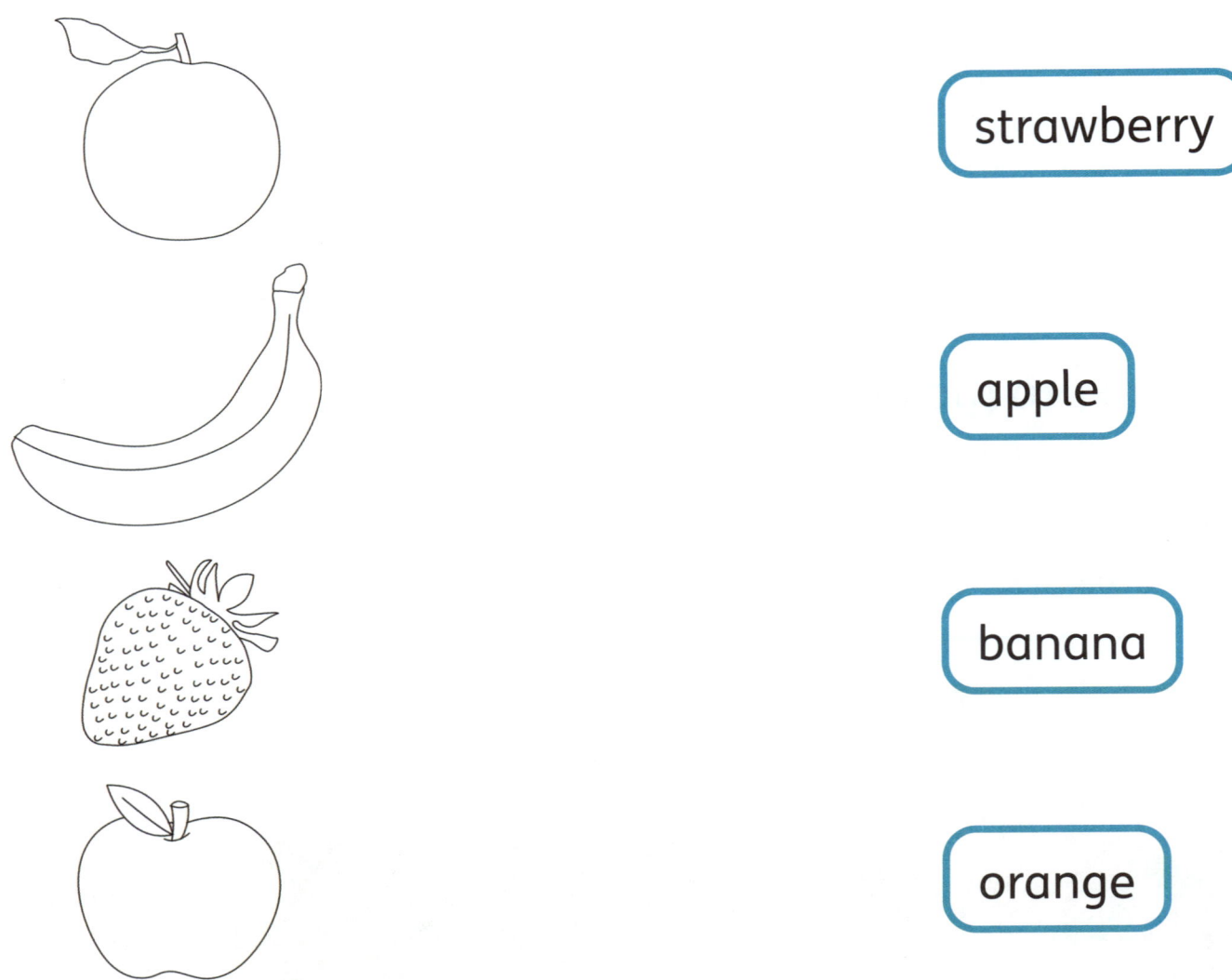 Show and tell.

Animals

1 Draw lines to match the adults to their babies.

_____ _____ _____

_____ _____ _____

2 Name the animals.

3 Tick the birds.

4 Count the legs of the adult animals. _____

1 Finish colouring the animals.

_____

_____

_____

2 Name the animals.

3 Count the legs. Write the numbers.

4 Show and tell.

Animal movement

1 Name the animals.

2 Count the legs.

3 How does each animal move? Draw a line from each animal to the correct words.

swim   walk   fly   slide

Animal places

1 Match the animals to where they live. Draw lines from the animals to the places.

2 Match the words to the places. Draw lines.

sea   jungle   grassland   arctic

3 What other animals live in each place?

Weather

1 Match the symbols to the words.

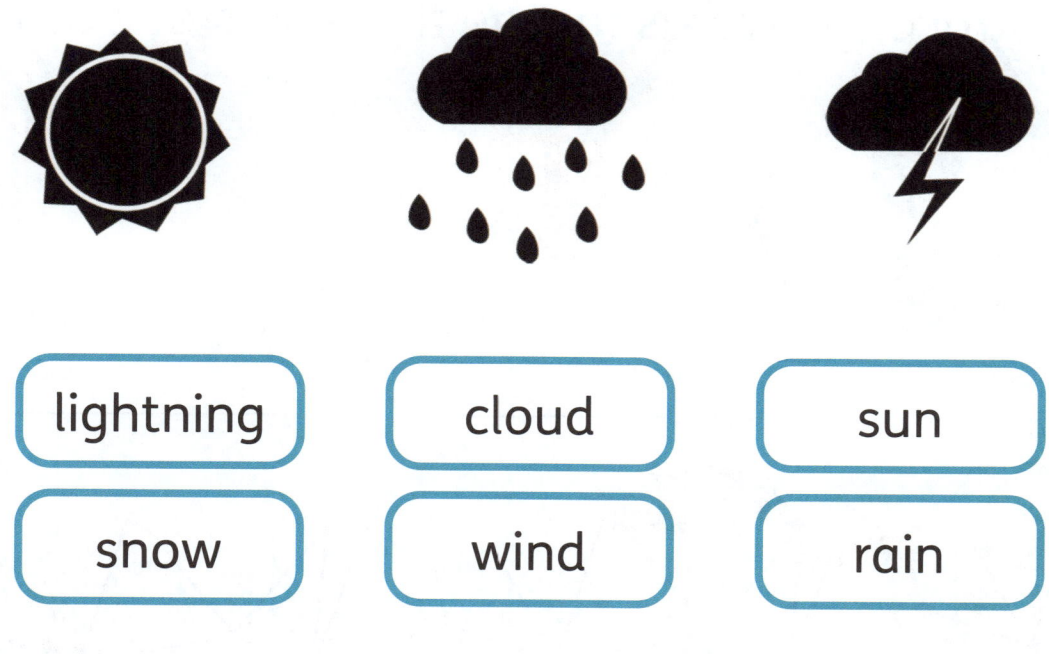

lightning   cloud   sun

snow   wind   rain

2 Draw the symbols for the weather today. Copy the words to label your picture.

Weather and clothes

1 Name the clothes.

2 The sets of clothes below are worn in different kinds of weather. Trace the words for the three sets of clothes.

cold weather

wet weather

hot weather

3 Draw yourself on a wet or a cold day.

Things which use the wind

1 Cross out the things which do not use the wind.

Types of weather

1 It's raining. Add rain drops to each picture.

river   lake   sea

2 It's snowing. Add snowflakes to each picture.

snow mountains

iceberg

3 Draw your favourite type of weather.

Show and tell.

1 Name the animals.

2 Cross out the animals which do not live in the soil.

3 Draw an animal in its underground home.
   Show and tell.

1 Look at the picture of a sweet potato. A sweet potato is a root.

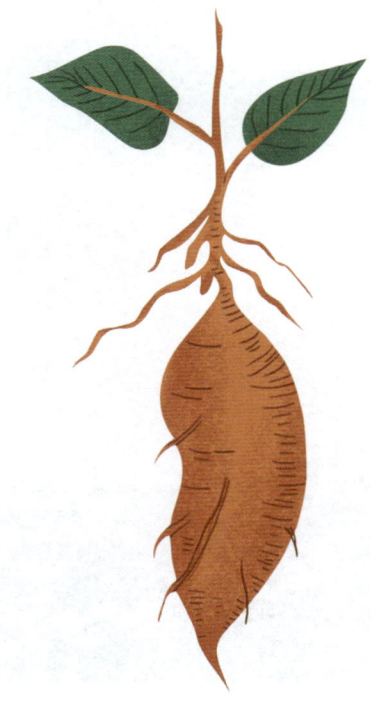

2 Draw the roots on these plants.

| carrot | radish | beetroot |

What seeds need

1 Look at the photo of seeds growing in soil.

2 What does a sunflower need to grow? Draw a sunflower and what you think it needs to grow. Show and tell.

1 Look at the wall patterns below.

2 Draw more of each pattern.

3 Use blocks to build wall patterns. Make a pattern with the different colours.

4 Draw a house. Show and tell.

5 Match the building types with the words.

house

school

hospital

6 Draw your building of the future. Show and tell.

Shops

1 Underline the names of all the **fruits** you can see.

| carrots | pears | pineapple | melons |

2 Count the number of each fruit. Write the number next to the name.

3 Draw your favourite fruit. Show and tell.

Food containers

1 Look at the food containers in the shop.

| _____ tins | _____ jars |

2 Draw circles around the sets of jars.

3 How many jars? Write the number.

4 Cross out the sets of tins.

5 How many tins? Write the number.

1 Look at the picture of a fish market.

| cod | crab | squid | clam |

2 Underline the names of those with shells.

3 Draw a fish.  Show and tell.

1 How do the things in the furniture shop feel?

2 Write H on the hard things.

3 Write S on the soft things.

4 Write W on the things made of wood.

5 Write F on the things made of fabric.

Museums

1 Write H on the things you hit.

2 Circle those things you blow.

3 Write S on the things you shake.

4 Underline those things you pluck.

hit	pluck	shake	blow 

5 Make a shaker with beads or seeds and a pot or a tin.

1 Put ☒ on the stone things.

2 Circle the metal things.

3 Draw a pattern.

4 Draw a pot or make a pot. Add your pattern.
   Show and tell.

5 Draw your own artwork in the empty frame.

fabric   wood

6 Make something for the gallery. Use fabric or wood or metal or paint.

7 Show and tell about what you have made.

UNIT 8

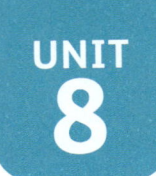

In the street

1 Write E on the things with an **engine**.

2 Circle the things with **wheels** that do **not** have an engine.

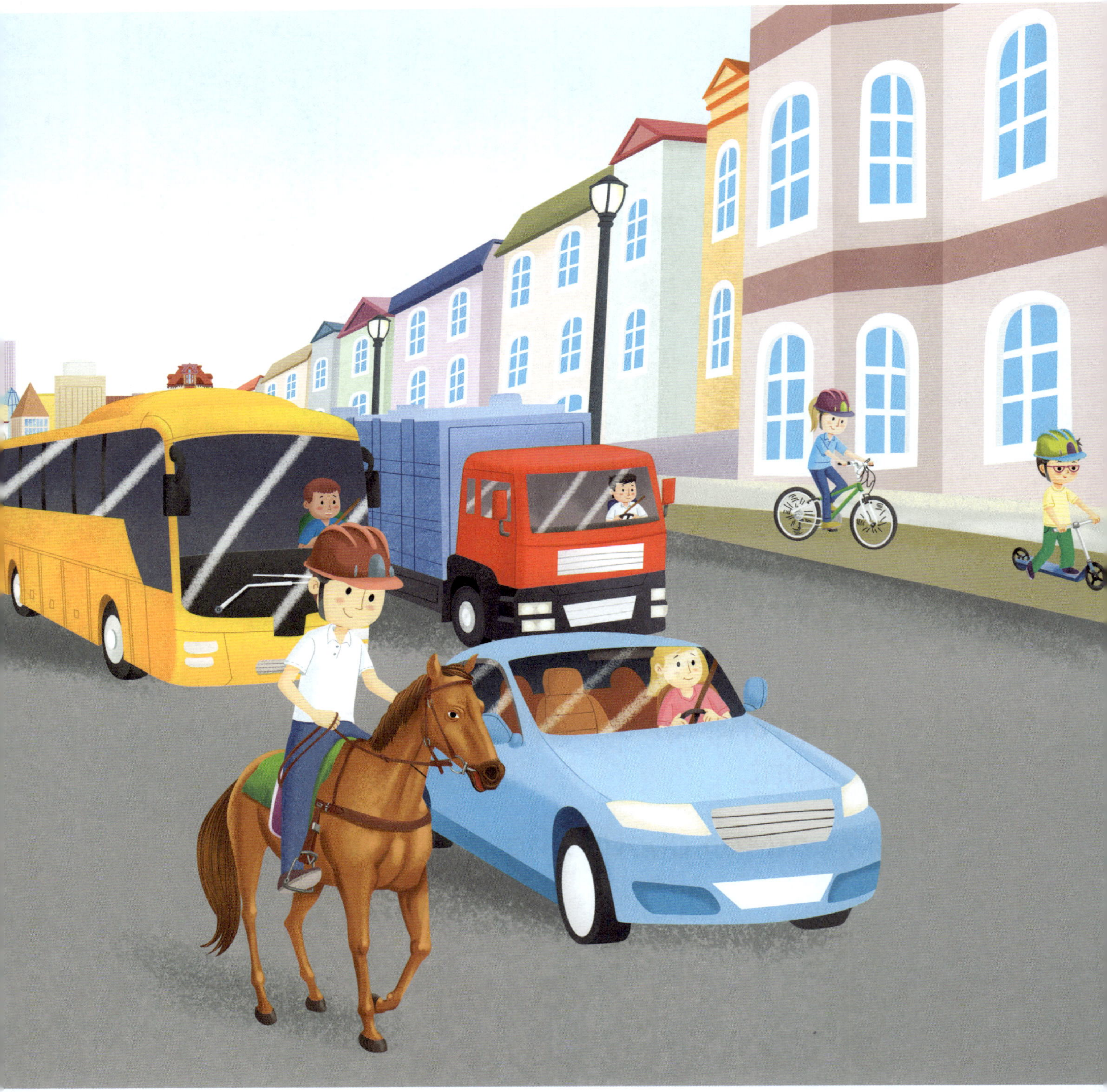

Sounds

1 Look for all the things making sounds. Draw circles around them. Count them.

2 Use your body to make 2 sounds.

3 Draw something that makes a sound.
 Show and tell.

Pull and push

1 Circle all the pulls. Count them. Write down the number.

2 Cross out all the pushes. Count them. Write down the number.

_____ pull   _____ push

3 Draw a large toy with wheels. Can you pull it? Can you push it? 👤 Show and tell.

In the park

1 Write up and down on each picture.

_____   _____   _____   _____

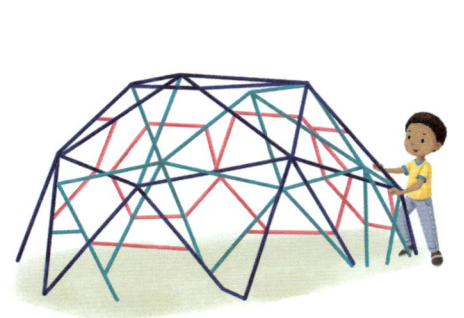  

_____   _____   _____

_____   _____   _____

2 Use a big book to make a slide. Slide things down it.

3 Draw what you have done. Show and tell.

Trees

1 Number the trees in order of height. Label the shortest '1' and the tallest '4'.

2 Draw a tree shorter than number 1.

3 Draw a tree taller than number 4.

4 Draw lines from the parts of the tree shown below to their names.

bark   leaf   bud

Water

1 Circle all the things that float.

2 Tick the things that swim.

3 Look for objects around the classroom. Find out whether they float using a bowl of water.

Movement

1 Circle the moving things. Draw lines to the words.

flying rolling running resting

kicking catching throwing

2 Use modelling clay to make two or more shapes that roll. Try them out on a slope.

UNIT 10 Staying safe

1 Tick things that keep people safe.

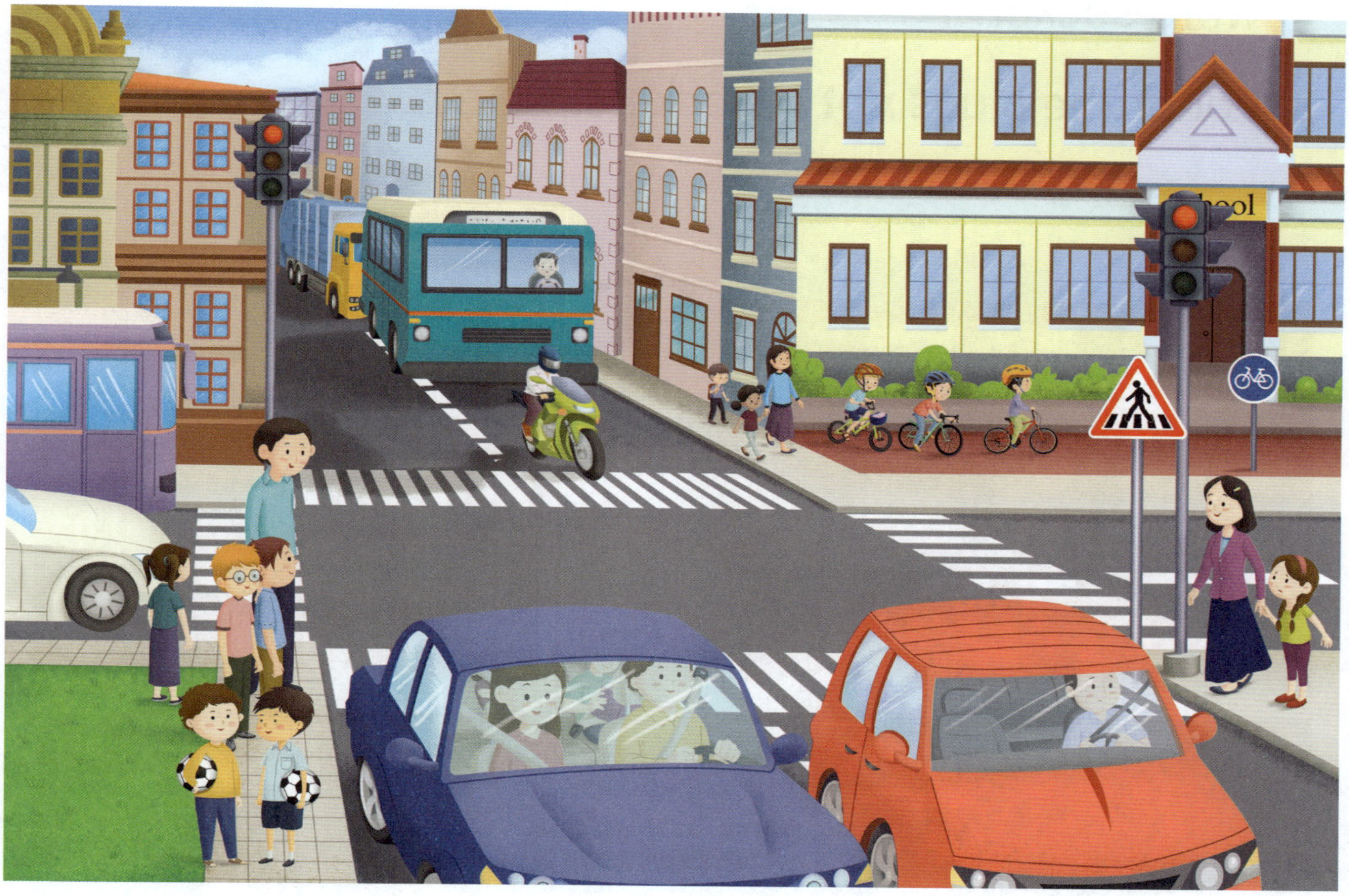

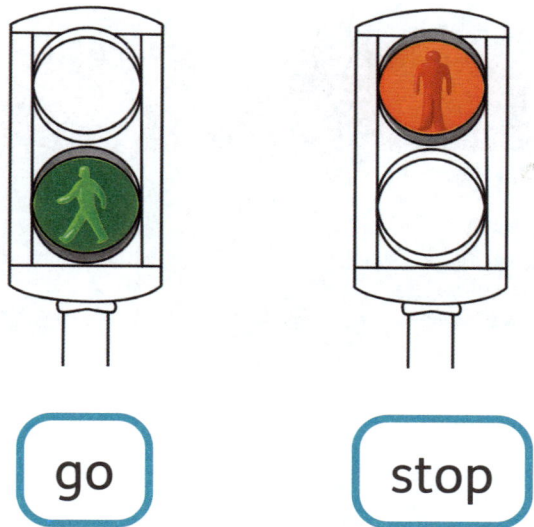

2 Draw lines from the symbols to the words that match them.

go

stop

Electricity

1 Put a ☒ on things that use electricity.

2 Circle the electric sockets. Draw lines from them to the word socket.

3 Count the plugs **not** in sockets. _____

socket

Fire

1 Mark all the flames with a ☒.

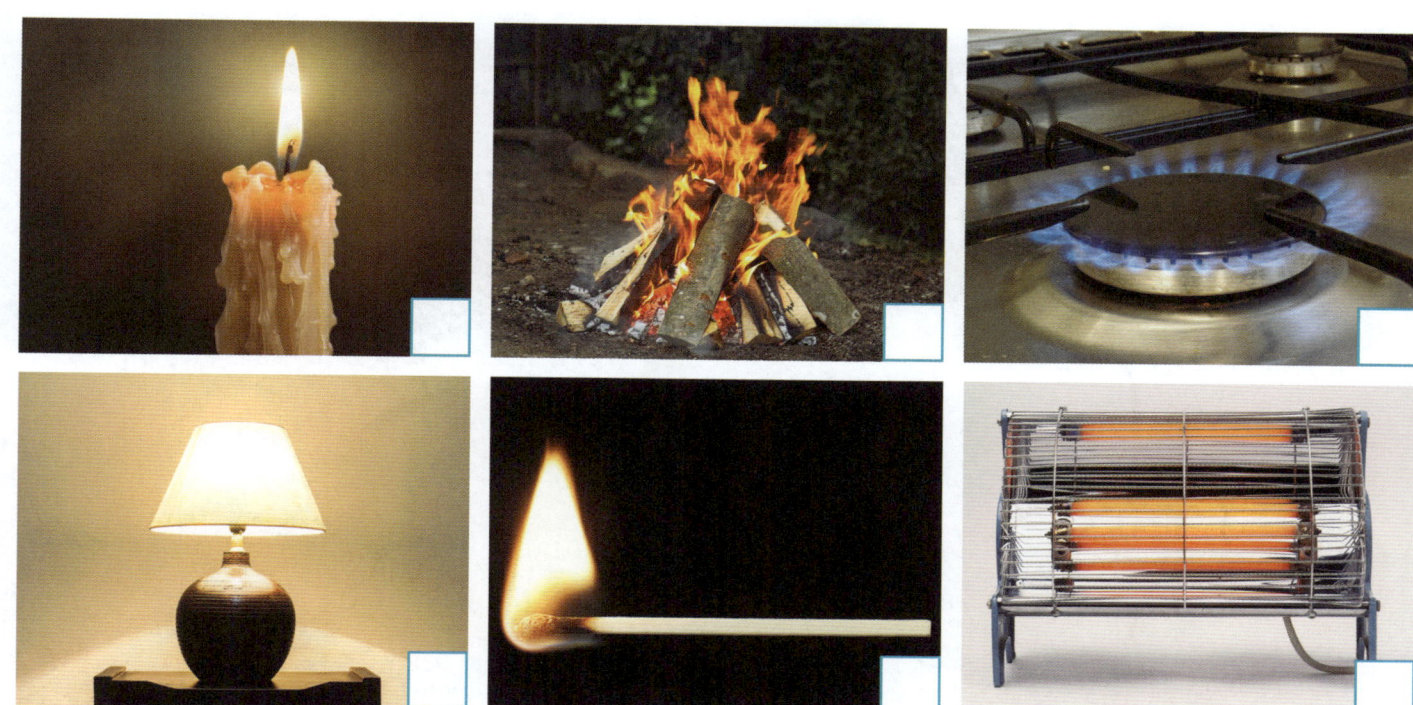

2 Circle the words that describe flames.

cold | burn | safe | hot | danger

3 Which things can you find in or near your school?

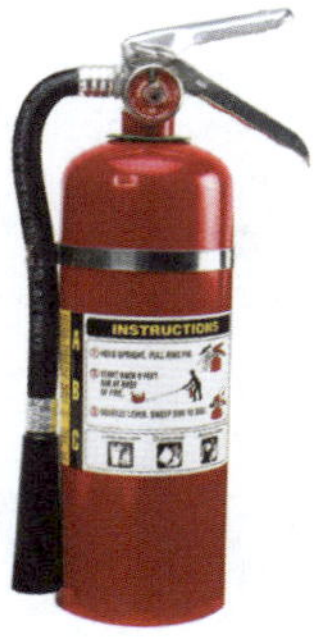

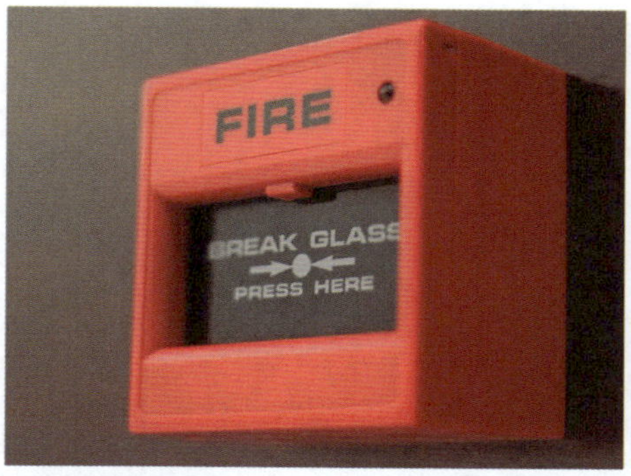

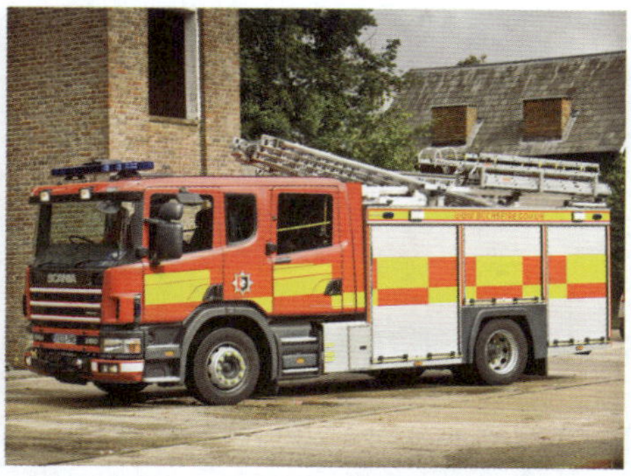
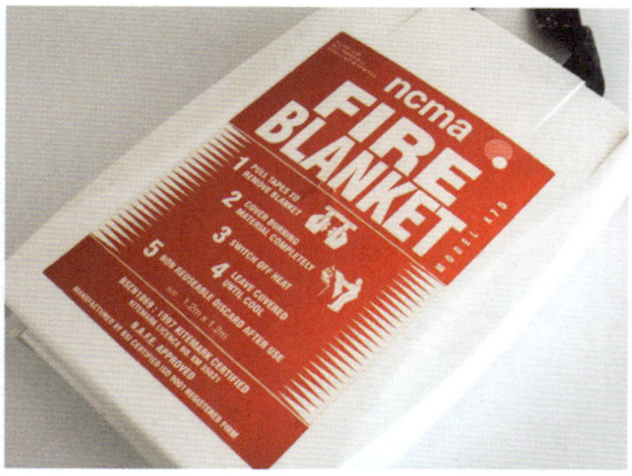

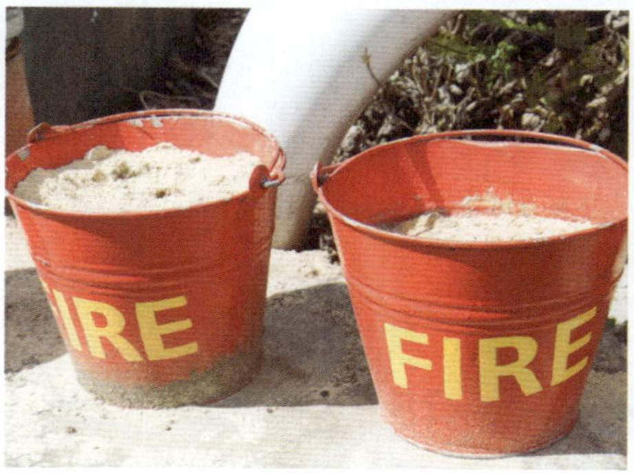

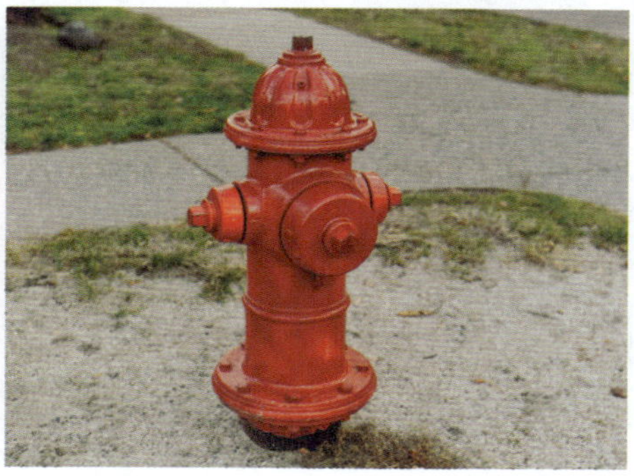

Danger

1 Draw a $\boxed{X}$ on the dangerous things.

2 Draw one more dangerous thing. Show and tell.

UNIT 11 Holidays

1 Draw lines to match the words to the pictures.

train   car   boat   plane   bus

2 Tick the ones you have used.

3 Circle the type of transport you like best.

4 How do the things in the pictures move? Mime the movements. Ask others to guess what you are miming.

Holiday activities

1 Tick the things you would like to do on holidays.

2 Match the words to the pictures. Draw lines.

float   slide   swim   ride

3 Draw your ideal holiday. Show and tell.

Glossary

burn – when an object burns it gives off heat. Flames rise from burning things.

describe – use words to tell what something is like.

electricity – electricity is a flow of energy. It usually comes from a battery or from the mains.

engine – engines are machines that can move things. Engines need fuel such as petrol, diesel, or coal to work.

flame – flames are produced when a fire is burning.

float – objects float in water when they are not resting on the bottom.

root – the part of the plant that takes water from the soil.

shell – shells are the hard outer coverings of snails and some sea animals.

soil – soil is a mixture of ground up rock and the remains of dead plants and animals.

sound – sound is caused by vibrations. You hear sound with your ears.